AF365505

Disclamer

The recommendations, advice, descriptions, and methods in this book are presented solely for educational purposes. The author and publisher assume no liability whatsoever for any loss or damage that results from the use of any material in this book. Use of material in this book is solely at the risk of the user.

W Formation: Fire Attack and Tactics
(Original title: Consejos para implementar la Formación W)
Copyright © 2011
José Musse
New York City, U.S.A.

Cover, illustrations and 3d art by José Musse.

Library of Congress Cataloging—in—Publication Data Available upon Request.

Dedication

To my wife Maria who has always supported me on each and every one of my projects, even the most absurd. Throughout the years tirelessly she has been responsible for all of the family details so I could spend hours writing and developing concepts as well as refining ideas for the benefit of all my brother firefighters.

With special gratitude to my friend Rocky Jean-Philippe, without his support this book wouldn't have been written.

To my children Carlos and Michelle for being an exhaustible source of joy and inspiration.

José Musse

Ackntowledgment

My gratitude and dedication of this work is to my team of 30 people who make *Desastres.org* and the *Waterbomber.com* possible.

Especially Alex Quintanilla, Ed Pearce, David T. Crews, Isabel McCurdy, Enrique Martín Cuervo, Fernando Bermejo Martín, Rony Iván Véliz, David Rodríguez Carrasco, Gerardo Fabián Crespo, Martín León Li, Eudo Hernández, Andrés Medina Villegas, Raúl Leiva Escudero and Mariana Sathya Llull.

To the firefighters who gave their photos to illustrate the ideas presented in this book.

Introduction

The **W formation** is one of the most elegant and beautiful maneuvers that firefighters have developed. It is so appealing to the eye that they're often implemented in exhibitions or drills, because they capture the attention of the audience like no other. For years I have had a fascination with this exercise and it has always caught my attention that there is not enough literature on this maneuver. It is surprising how little literature exists on the subject, especially because it is an essential maneuver to combat fierce fires in flammable liquids and gasses.

Over the years the **W formation** is executed by firefighters several times, however, I noticed that when displaying the fire with two bright shields forming mists of water, it rarely follows a standard.

In a regular training exercise firefighters are trained in their execution, but rarely put emphasis on certain details. For example, often there isn't enough time spent explaining how a team should be lead in conducting a **W formation**, especially when visibility is almost zero and when water touches the flame, conditions only worsen.

I cannot imagine a more dangerous scenario to move forward blindly above flammable liquids.

In this manual I want to suggest some standards, propose some principals and concepts. I hope they are useful.

José Musse

5

Index

The most interesting part of this maneuver is that it is similar to running a ballet on stage. It is not a —*Solo*— therefore requires that firefighters taking part be skilled in its execution.

W Formation: Fire Attack and Tactics

Photo: José Musse.

The first step is to familiarize yourself with the equipment which you are going to use, in this case hoses and nozzles of water to be a natural extension of the body of each of the team members that develop the **W formation**. The most interesting part of this maneuver is that it is similar to running a ballet on stage. It is not a —*Solo*— therefore requires that firefighters taking part be skilled in its execution. This maneuver requires a good nozzle. The practical way to find out the status of a nozzle is reviewing the types of mists and their qualities. An uneven wide-angle fog should not be used. Firefighters have the dispersion of water quality to

<table>
<tr>
<td>

WARNING

Basic knowledge of hydraulics is needed. Make sure you have the necessary knowledge.

</td>
<td>

IDEAS

The maneuvers are highly theatrical, almost choreographed.

</td>
<td>

TIPS

Know every technical detail of your hoses, pumps and nozzles.

</td>
</tr>
</table>

succeed in their operations. Regular maintenance should be done to ensure an optimal outcome. Get in touch with the manufacturer or representative of the brand if you are unsure.

For the officer that will lead the **W formation**, there are other additional requirements as solid knowledge in fire ground hydraulics. Beyond this theory you must be skilled to recognized, when a jet is not effective.

You can calculate pressure and verify that adequate pump pressure is correct. Additionally, it is expected that a fire officer easily knows to recognize smoke from water vapor.

The latter means that the extinction is being successful; the first is an indicator of problems.

Beyond the knowledge expected of every officer, it must be right to run the task that is expected to comply with the **W formation**. This means you have solid knowledge in this maneuver. Nobody wants to be led by a novice. This need for experience covers a variety of tasks ranging from putting out a fire to pressure, closing a valve, to extinguishing a spill of flammable liquid, etc.

The key to success is not only knowing more. Of course, knowledge helps, but this task requires many hours of practice to make it safer and more efficient. A **W formation** is a step back to a firefighter that can expect after

WARNING	IDEAS	TIPS
There are things you should probably know before operating hoses in industrial plants	Practice and test your fire equipment according to NFPA 1410.	Pressure fire is all it takes sometimes, as a good fire extinguisher to do the job.

having a complete mastery of work as a nozzleman. The firefighter after proceeding with this maneuver should be skilled in assembly and handling of hoses of different diameters. The time and ability it takes to maneuver as a team or individual working with hoses is important.

The NFPA 1410 Standard on Training for Initial Emergency Scenes Operation is an excellent reference, which could and should be used before improving this maneuver; it guarantees a minimum compression and ability.

Moving and walking with hoses make it easier for firefighters to go beyond the safer pattern of the water mists.

W Formation: Fire Attack and Tactics

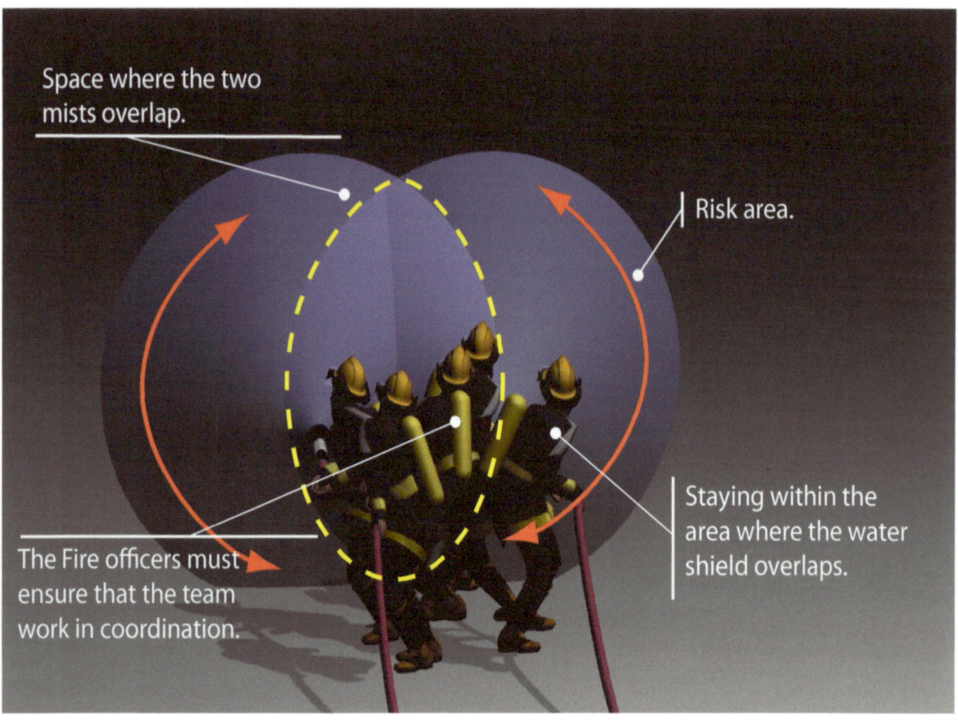

To develop an adequate **W formation**, it is necessary to understand the anatomy and dynamics of water mists. It takes many hours of practice to get the team of firefighters to spend most of the time within the safer area, precisely where the two shields are located. Moving and walking with hoses make it easier for firefighters to go beyond the safer pattern of the water mists. In the haze of the protection there is an area that is extremely sensitive and vulneable.

WARNING	IDEAS	TIPS
In violent intense fires avoid separating from the group.	Be familiar with the main industries in the area of influence in the unit.	Always keep track of the time duration of your air. Someone outside should take that same control.

José Musse

The fog is just less dense in this area of weaker water, which makes it vulnerable to strong winds and pressure fire. It is in this place that the phenomenon brings debris vortices that radiate heat to your team.

When the *W formation* is not supervised and directed by an outside official and all liability rests with the leader of the group that runs the same formation, often slightly separates the team just to see what is happening, that puts him at an unnecessary risk. Propper coordination and management of the scene never allows firefighters to enter a zone of risk improperly.

W Formation: Fire Attack and Tactics

Firefighters must test their nozzles before entering problem areas. Notice in the photo, fog stream is uniform.

Photo: Charles A. Edwards Jr., US Navy.

Narrow fog (15°— 45°) is ideal when you want to focus the power of water in a specific area.

Photo: Barry Ril, US Navy.

WARNING
Policy should be an adequate preventive maintenance of all firefighting equipment.

IDEAS
Check your nozzles and the different jets on a monthly basis. Observe the difference between them.

TIPS
The best firefighter is the one that never suffers injuries or wounds.

José Musse

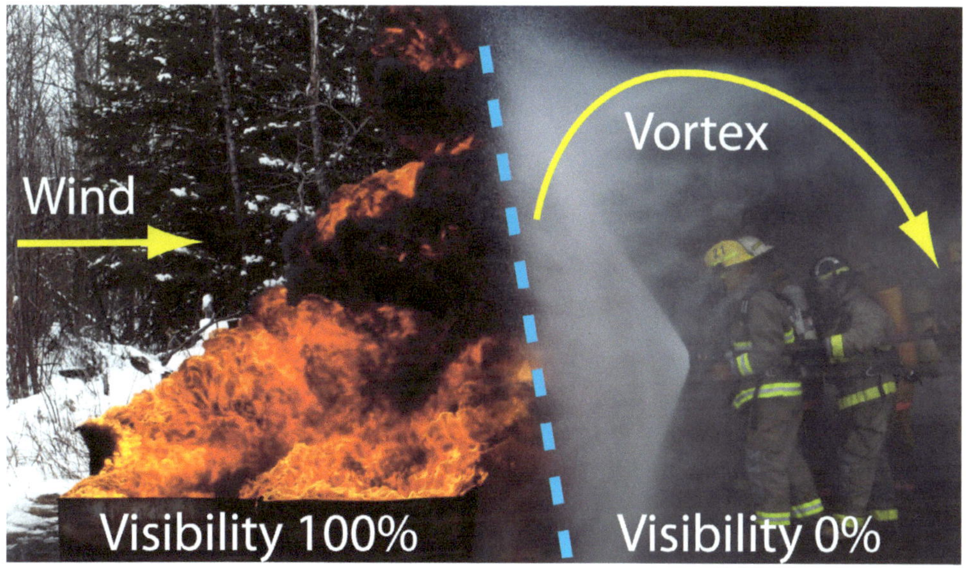

Photo: Chief Spencer del Conmee Fire Department, Canada.

When the *W formation* comes in contact with fire you can see the thermal impact caused. Notice the above photo, a team of firefighters in compact formation, shortening the space between each other. You should be very careful at this stage, any violent fires that could emerge or any interruption to the flow of the water, including a reduction in flow or significant pressure that would put the entire team of firefighters in immediate danger.

The image can be seen and contrast is visible, on one side is standard and the other conditions are zero visibility. Conditions in the picture look better than they are. Firefighters feed cold air from their SCBA and may suffer from

WARNING	IDEAS	TIPS
Let your team know if you have lost visibility, maybe it's not the same for the rest of the group.	Practice with zero visibility; learn to be guided by instincts. Hear and feel the movements.	If only the nozzlemen loose visibility he should change positions with the assistant nozzleman.

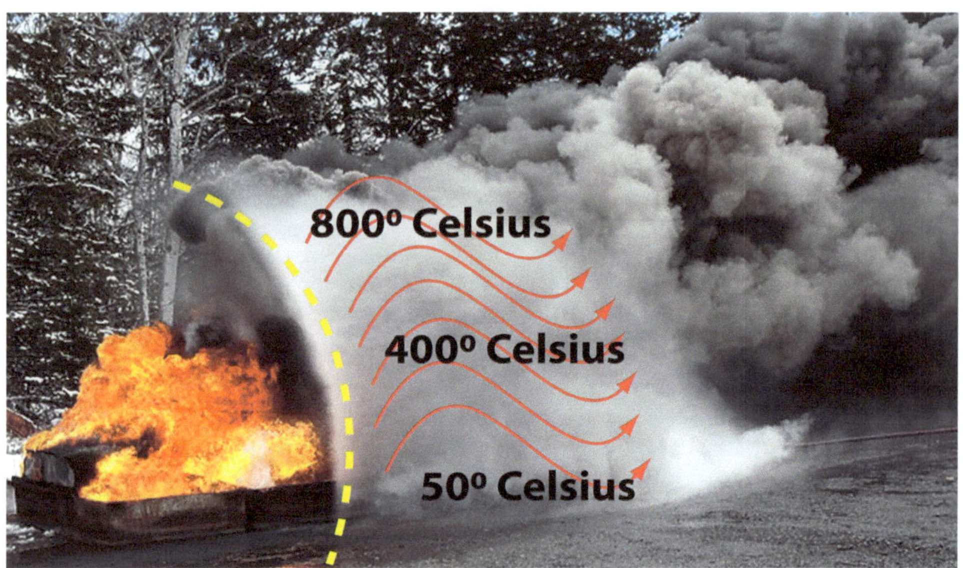

Photo: Chief Spencer del Conmee Fire Department, Canadá.

condensation of shock temperatures inside their mask, making it extremely dangerous. Considering that not all is breathable atmosphere risk is greater.

To which we must add the fine droplets of mist falling on the exterior of his visor worsening the whole scenario. Although it may be difficult for firefighters to see each other, they cannot see anything in front of them, precisely where the danger lies. Hence the need for external supervision and adequate radio communication to avoid falls in puddles of fuel or other hidden dangers in view of the team developing the *W formation*. In interior fires the visibility problems are fought in two ways, ventilating and lighting.

José Musse

Photo: José Musse, Engine 308 FDNY.

When fighting fires outside, visibility is usually lost and there is not much that can be done.

In real life scenarios all resources must be exhausted to combat these fires with the wind. It is recommended that this fire team maneuver is prepared to go back quickly if a fault occurs. Practice is needed until this maneuver is done perfect. An alternative in this maneuver is to reduce the exposure of the body by ducking. The downside to stooping to this difficult maneuver is an early withdrawal if necessary.

WARNING	IDEAS	TIPS
If the heat is too intense for you despite the fog, let the leader know don't feel ashamed.	Prepare a pool of fuel to burn. Observe its evolution from start to finish.	Learn how to control your breathing with SCBA. You will achieve more productivity at the scene.

However, reduced body exposure to fire firefighters protect them including serving as shields in case of interruption of water flow, the nozzle team will be in the lower area of land that is usually the coldest in a fire. Note that this team is acting against the wind. Training in these conditions is desirable, but only to teach the team the real difficulties that can be found, in addition to understand because it is necessary to have a high mobility as a team. In many scenarios the direction and intensity of the wind can change suddenly.

In this kind of fire there is a stratification of temperature, but will only be noticeable in its vicinity. This team will not succeed in its mission or will achieve with great difficulty. The downside of working in adverse conditions is that they consume more resources. Water and air (SCBA) as the team could spend more time in extinction.

Therefore urged to ensure provision of a stable flow and pressure, that is not going to fail just as the front men need it most.

Photo: Chief Mike Norman, Amber Fire Department, Oklahoma.

Recommendations on Implementing the
W Formation

✓ Before acting have a clear action plan.
✓ Listen more, talk less.
✓ The principle of maximum security is; if you take one step forward is because you know how to take two steps backwards.
✓ You should know what is expected of you and your team. What objectives you need to achieve and the approximate time it will take.

WARNING	IDEAS	TIPS
Pump pressure and water reservoir is key before starting an offensive action.	Planning is important, it offers security and allows you to evaluate progress.	A good plan starts with simple questions. What can be done and how can it be done.

✓ Have a maximum deployment time, a maximum time of action and a maximum time of withdrawal. This way you will not experience lack of air (SCBA).

✓ Keep an eye on the team developing the operation; you cannot be distracted every minute counts. Fires in flammable liquids are violent and sometimes unpredictable.

✓ Have a Rapid Intervention Team (RIT).

✓ See if it is necessary to implement an Away Team (AT) see the Away Team chapter (Page 26).

✓ It is important to consider that excessive operating hoses do not affect the overall development losing all or some of the pressure or flow. You should know what is expected of you and your team.

✓ Objectives are needed.

José Musse

Photo: Firefighter Brian Rourke, New South Wales, Australia.

The Alternative Supervisor

When there is no fire officer that can monitor from afar (binoculars) the path that the nozzle team should follow warning of hidden hazards or other situations, we can establish another line of hose nearby, but from another position with the objective to have another visual perspective. The water line has no role in the extinction, it should be under the operation of an officer whose mission is to assist the *W formation*, monitoring the evolution of the fire and providing information to the leader of the *W formation*,

Photo: Michael Meadows.

which in turn orders his men the best course of action. In this case the risk that is involved with this fire officer whose job is to supervise firefighting tasks looses sight of his post. If this occurs it can compromise the safety of the entire team developing the *W formation*. There should be other officers as backup in different positions around the fireground. It is the Incident Commanders obligation to explain and define what is expected of each position. Evaluating all firefighters in the field and making necessary adjustments to the plans according to what you want.

<table>
<tr><td>**WARNING**</td><td>In case you cannot extinguish the fire and need to leave, do not close the nozzle, it could burn you.</td><td>**IDEAS**</td><td>Practice the "Duck Walk" it is the safest way to move in structural fires.</td><td>**TIPS**</td><td>It is important to maintain hydrated after performing actions of this type.</td></tr>
</table>

It is important for the team before acting on the ground to have a mental road map the way forward, the obstacles and problems faced.

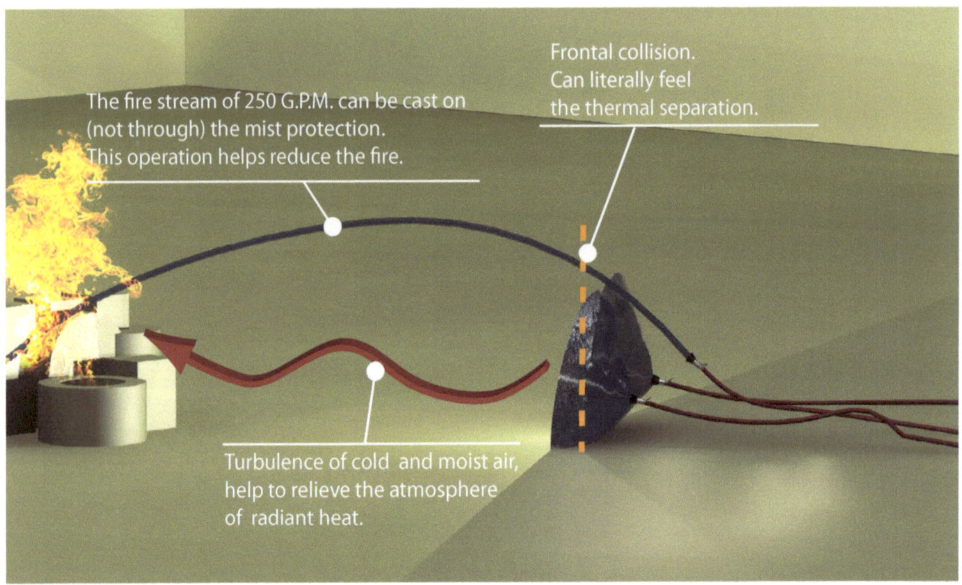

When I was in command fortunately there were no casualties or injuries. One of my priorities has always been to promote tactical defensive maneuvers. However, the **W *formation*** is by nature an offensive maneuver which is performed in an open environment saturated with heat and gases. Therefore, we must double our efforts to not send firefighters unless extremely necessary. Look at other possibilities, such as saturating the environment with water from monitors or have additional hoses deployed to protect the team that is coming.

WARNING

Visibility is always a challenge. Don't underestimate the protective mist even if you are working in the light of day.

IDEAS

Train as often as possible, always with SCBA and obstacles.

TIPS

Have two officers with binoculars and radios to monitor your team.

Look for the more experienced for this task. It does not work like in the movies when you ask for volunteers, you know your men and you know who is best for the job.

Field Plans

It is important for the team before acting on the ground to have a mental road map the way forward, the obstacles and problems faced. It is imperative that this task not be over looked by the Incident Commander. A previous meeting with the team to review the work or assignment is required. There is no better way to train firefighters for the *W formation* which train to overcome obstacles and walk on slippery areas. Do not forget that often these incidents involve liquids or oils, which are also in an area covered with fine water droplets.

A variant to the *W formation* would be to launch a strong jet of water from behind; if it does not lead us to extinction at least it will appease the flames.

While the mists of the protective shield protects against radiation as firefighters approach the fire and gas.

Relief & Reservations

Be sure to estimate the time it takes for the team to run through the area until the site works final extinction, possibly by closing valves for gas or flammable liquids that may fall short of compressed air (SCBA). Although fires of flammable liquids or gases that are battled outdoors usually are breathable. It is highly recommended to operate with SCBA.

They are still toxic and dangerous, especially if you breathe in flammable gases that are waiting to deflagrate. In this case if the team runs the risk of running out of air

WARNING
Discuss with your team in advance what you need to do and how.

IDEAS
Explore the Away Team concept in other scenarios especially the big scale ones.

TIPS
Practice this maneuver two hours weekly for six months.

work on an Away Team. Do not bother looking for the literature on this term that applies to the firefighting field, it's my invention. I borrowed this term from the Star Trek television series and used it to refer to equipment that will fulfill missions support. Not to be confused with the Rapid Intervention Team (RIT). As I explained above, if the team runs out of air we can declare Failsafe and send an (RIT).

However, if we anticipate that to happen because there is no emergency declared in what may be planned?

In the case of a group advancing to a location where it will close a valve that is leaking toxic and flammable gases, we must then send the Away Team. Not to assist directly in the operation of control, but specifically planting air tanks in the field.

When the team is sent to close the gas valves or perform other tasks they should know where and how many bottles (SCBA) await them.

Of course there should be a proper plan that marks the location of the support material, previously agreed with the team.

José Musse

Estimate the necessary provisions that the **W formation** may need and make sure the Away Team will provide them with what's needed for their return. The funds allocated are planted in the field becoming a pocket of resources in a mini staging area. In case of a failure in a SCBA or a premature exhaustion, the team of firefighters will know that very close to them is a source and must not go all the way back. An Away Team can provide them with extra tools needed to complete the task with shears, sledgehammers, etc.

Think about the Away Team, how the firefighters prepare the logistics of the scene.

In industrial incidents the equipment planted by the Away Team perhaps could be supplied by the company.

Always coordinate with the company's security department.

In certain situations it is better that your incoming Away Team be first on the scene, perhaps covered by the initial *W formation*, only putting extra resources at the scene, then another *W formation* to do the final work of extinguishing or closing valves. Although it is not necessary to create a *W formation* for an Away Team if the resources are not placed in an area where there is excessive radiant heat it would probably no require to extend extra hoses.

The Incident Commander should determine what equipment is needed to take to the location. Planning and proper execution is key to the success of the operations in the line of fire.

José Musse

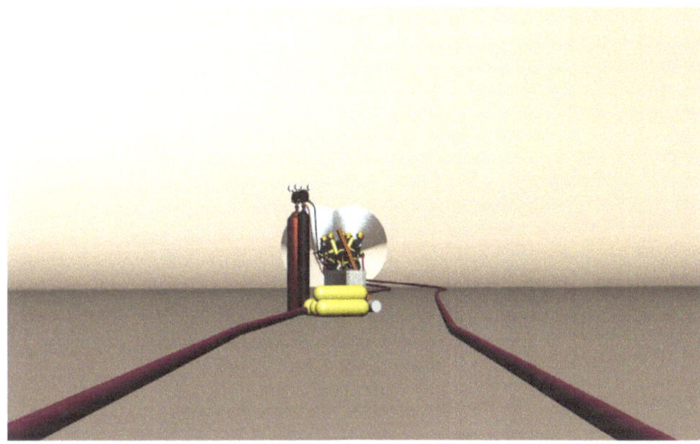

The distance between the fire and the resources sought by the Away team can vary from one situation to another. Depending on the equipment to be moved and their sensitivity to hot environment.

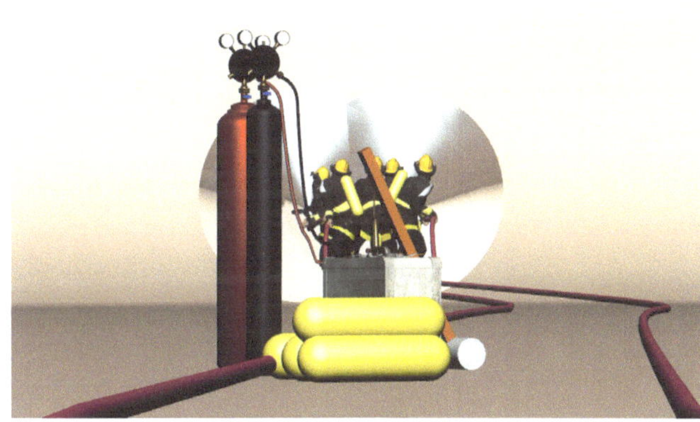

As close as possible to where they are needed, be sure to be the first that could determine the location of these resources.

WARNING	IDEAS	TIPS
Do not place equipment or tools just to place them on the ground. Must respond to a general plan.	If you have air cylinders consider protecting the valves. Flammable fires produce large quantities of soot.	The quality and the good condition of the boot prevent slips and falls.

Training

It is important that firefighters individually or as part of a team to maneuver and promptly respond to any situation.

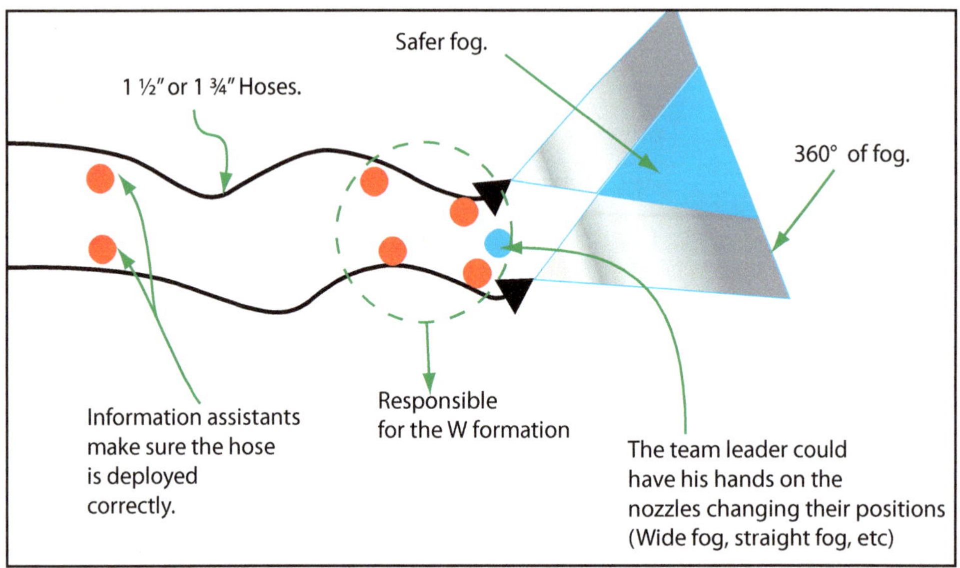

In my experience to develop an appropriate **W formation** may take six months of continuous practice devoting two hours a day minimum. Problems arise when the team of trained firefighters which is not the same with which you respond to an emergency where the training should be deployed. This is evidence that individual solutions do not help overall if not properly implemented. In this case it must be fire service policy to obtain expertise in this area.

Do not be afraid to consider running the hoses from the **W formation** with 2½". The well trained team of firefighters learns how to achieve mobility, take time to understand how to quickly move heavy hoses.   Firefighters should probably include more assistants if the length of hose is long, but finally achieved if you persevere. However, do not use a single length of hose for each line displayed in any exercise. The correct number is 4 to get some mobility in hoses that are 1½". If the distances require greater lengths of hoses they should connect to hoses that are 2½" which is preferred to keep the desired flow, about 125 G.PM. (575 L.P.M.) If implemented weekly as a regular exercise firefighters performances should improve considerably.

Considerations When Training in the *W Formation*

✓ Never do it without full personnel protective equipment (SCBA). If your service does not permit you to spend on air refills then do not connect, but if you must use it be sure to wear it to get used to the weight and discomfort of the equipment.
✓ It is important that firefighters individually or as part of a team to maneuver and promptly respond to any situation.
✓ Each exercise is preformed to add a new degree of difficulty. Work with obstacles and uneven surfaces, include debris if possible. After an explosion you can find lots of debris.
✓ Learn how to synchronize your movements; decrease the rate of progress as well as bending or standing must be sequential.
✓ Learn how to communicate with signals, it will save air.

W Formation: Fire Attack and Tactics

✓ Deploy the *W formation* day and night.

✓ When making the *W formation* at night give clear indications to the lighting team to intentionally blind the firefighters.

✓ Reduce visual visibility to all firefighters, using plastics in the visors or other devices.

✓ Practice advances and withdrawals, withdrawals must be particularly fast.

✓ Have your team react swiftly to the loss of flow or pressure.

✓ Know what skills and tools are needed to control fires in gate valves.

✓ Implement an oblique circuit so that your team can tour and deploy several sections of hose continuously adjusted. Put obstacles and slopes; try to simulate a scenario of rubble.

José Musse

The Rule of Three 7's

In most situations when working with the *W formation* the team of firefighters manages to extinguish in a few minutes. In some cases it is so easy that it doesn't require further action. However, it may happen that the situations become more complex, especially when incidents involving petrochemical plants or ARFF events. In some cases, firefighters should make pathways which are not short and will take time by the amount of debris or fragments found on

WARNING

Pressure fires in gate valves are particularly complex.

IDEAS

Examine past cases to learn from successes and errors committed.

TIPS

The Three 7's rule could be used in any incident.

the ground. In this circumstance you can use the rule of three 7's, this rule is very simple and helps keep in perspective the work of SCBA. It is estimated, 7 minutes of deployment, 7 minutes of work in front of the fire and 7 minutes of withdrawal.

Of course there will be many incidents that do not require a walk of 7 minutes from cold to hot zone.

Fortunately, the standard is a core objective. That on 20 minutes firefighters are warned that they should initiate the return and on 21 minutes they must be in place.

It also sets a good margin of safety if we consider 30 minutes of air in the tank.

José Musse

Working within the hoses provides better alignment, which will impact on better progress.

There should be no distractions in your team, it is important to hear only one voice that of the leader.

W Formation: Fire Attack and Tactics

Practice with two lines and a length of one or two sections of 1½" hose on each side. On the ground you want to have expertise in this maneuver. There is no effort; it does not require work for more coordination. You can start the first day with maneuvers, but not there after. One of the greatest challenges for implementing these exercises is to work with obstacles. Fire safety depends on the integrity of the mist, but often in settings where it is to work more efficiently, in this scheme there are objects that obstruct the passage or distort the fog. Some of the most interesting exercises I ran were to close a damaged valve, which is accessible walking through several 55 gallon cylinders, some with fire.

WARNING
Be ready to act if a member of your team is injured.

IDEAS
Practice as if you were to withdraw and later if one or two members of the *W formation* are disabled.

TIPS
An Away Team can leave a stretcher close by to expedite the evacuation of a fallen firefighter.

José Musse

One of the major inconveniences is time; the firefighters have close to 7 minutes to do the job, the time it takes to reach the valve and what they need to leave the danger zone is key. (See The Three 7's rule at page 39) Training should always be done with counting time, so you could see the improvement of the team. Training must consist of one member of the teams *W formation* suffering an accident and shall be disqualified and removed without cancelling the two wide fogs. May be interrupted, but must be replaced and the reaction of one nozzleman to the fallen nozzleman is key to insuring the integrity of the group.

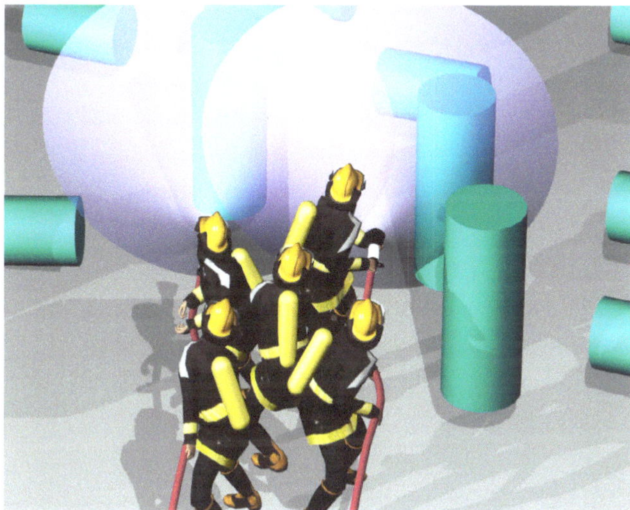

Add big and heavy obstacles, that are not easily moved which forces the team to work around.

Then put up a path where the firefighting equipment must necessarily adjust their laying hoses. Use at least four segments in length and provide them with assistants if necessary. Gather obstacles and complex paths. Although the *W formation* takes place in open spaces it is important to understand that buildings can end up semi demolished by a gas explosion.

WARNING

Balance of movement and prevent synchronizations is key for a *W formation*.

IDEAS

Become trained in closing valves and blocking pipe lines in poor condition.

TIPS

Always use full personal protective equipment.

They should be aware that the group might require to retreat from a violent fire and should not be so close to the nozzleman to become an obstacle and then fall to team members in full execution of the maneuver.

He who is the leader of the group decides the direction of the jets and ensures that the position is adequate; he also makes the necessary corrections to ensure maximum efficiency. Should keep the pace of progress, must also check at any time that the group is compact, giving instructions signaled with his right hand up most of the time so you do not need to speak saving air.

WARNING
A good firefighter must recognize when a fire stream is not being effective.

IDEAS
If the *W formation* has compromised its visibility outside, an external officer must inform the status of the fire.

TIPS
The leader, to ensure that its formation is compact can extend his arms to help in loading the hoses.

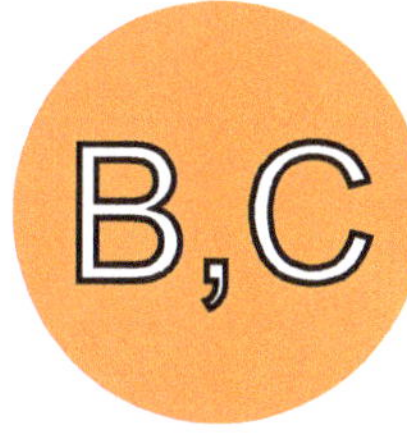

The nozzlemen of the group should respond synchronously. Unevenness in their fire streams would cause the whole group to be at risk. When a group leader orders a pan, it should start from left to right. It happens, many times the group is not synchronized and nozzleman B could start panning to the left while nozzleman C starts at the right creating a thermal divide that affects and seriously jeopardizes the group. The nozzleman C paces and executes speed, nozzleman B matches your brand.

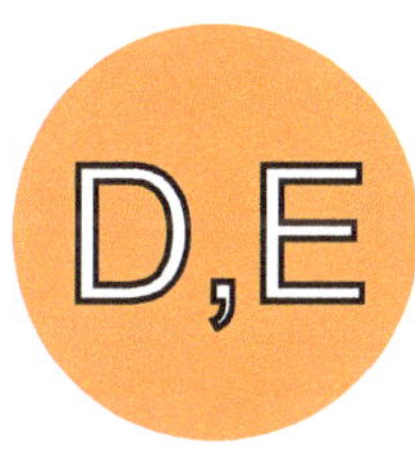

They are responsible for assisting the nozzlemen, who do not have to wrestle with the hose which is stalled. Space management is key, as close as an extended arm that touches the shoulder of the nozzleman or use the weight of your body when it should be kept more compact, yet it can leave the training to ensure they have the necessary hose extension.

They should be aware that the group might require to retreat from a violent fire and should not be so close to the nozzleman to become an obstacle and then fall to team members in full execution of the maneuver. The **W Formation** can be supported with solid streams and master streams from other positions. It seeks to further lower the temperature by reducing the intensity of the fire.

WARNING	IDEAS	TIPS
If one or two monitors can do the **W formation** job, do not implement.	Additional hose lines can protect the equipment in case it's needed.	The Away team should never leave unprotected sensitive material in dangerous environments.

The Incident Commander may establish more men to help on long journeys which will help with heavy lengths of hose.

This way the assisting nozzleman should not leave the group temporarily or less often. These firefighters did not need to integrate or be considered as part of the *W formation*, only the support as any other firefighter on the scene. If an accident occurs or fails its jets can assist the firefighters that develop the *W formation*.

By 1988 I created a basic sign language for communicating among firefighters that has evolved since that year and have been teaching with great success.

W Formation: Fire Attack and Tactics

Air is the most precious value we have with us when fighting fires or responding to an incident involving hazardous materials. When I started as a firefighter, in our Engine we only had four SCBA and two extra cylinders. We were forced to be efficient if we wanted to do well and survive together.

Necessity is the mother of invention; one day when I was watching a war movie on TV I observed how the soldiers communicated with each other in sign language. I asked myself, we as firefighters should have our own sign language that meets our needs.

Protective masks have done a great job hindering the natural flow and tone of our voices, making it more unintelligible when we need clarity. By 1988 I created a basic sign language for communicating among firefighters that has evolved since that year and have been teaching with great success. For the teams that perform the **W formation** is particularly important as saving air means more productive work. It isn't my intention to reinvent the wheel, most of the sign language used is already known.

When working with the **W formation** it is a challenge to make the signs visible for the whole team. In some cases you can just raise your hand as high as you can to make the prompts. They should not be very fast, while the team should be positioned for viewing. Some signs are specific for the nozzle team indicating blasting, panning and others, sufficient to do so at their eye level in front of them more or less where the nozzles are attached.

José Musse

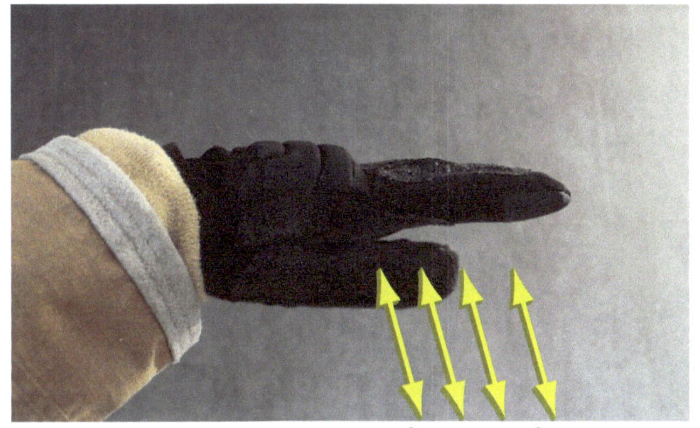

Squat

This sign in particular implies walking like a duck, which is the safest way to move carrying heavy hoses.

Photo: Carlos Carasas.

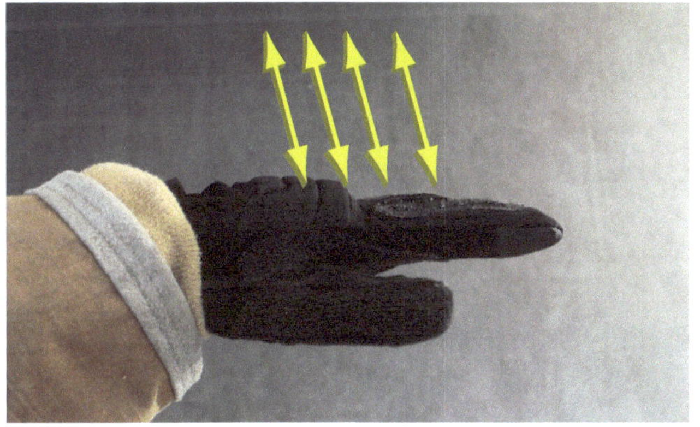

Rise

This sign is the same as the previous one with an upward emphasis.

Photo: Carlos Carasas.

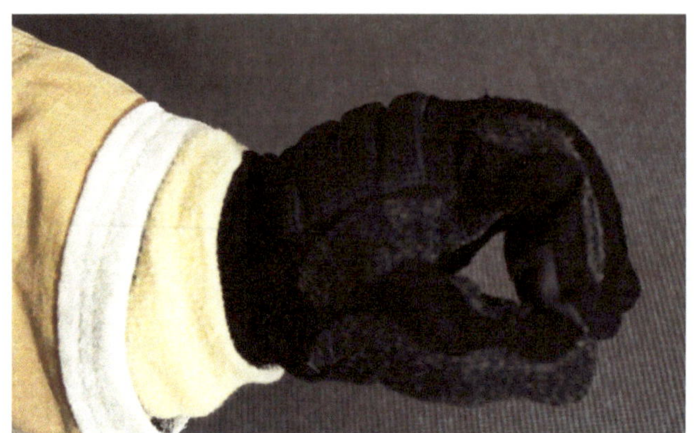

Photo: Carlos Carasas.

Hose
This sign indicates hose, which requires an additional line. This signal will be followed with another one indicating the size of the required hose.

Photo: Carlos Carasas.

1½" Hose
This sign indicates a 1½" hose.

Photo: Carlos Carasas.

2½" Hose

This sign indicates a 2½" hose.

Photo: Carlos Carasas.

Time

A universal sign where the hand forms a "T", in most cases it is used to indicate that the fire crew's equipment is running out of time and need to be removed for having low air supply (SCBA).

Photo: Carlos Carasas.

Rope

This sign simulates a central hole as if you're holding an invisible rope.

Photo: Carlos Carasas.

Hurry Up

This sign indicates the hands are performing a rotation which means to speed up the march.

José Musse

Photo: Carlos Carasas.

Hose Panning

This sign indicates commands that the hoses must make a panning or scanning from right to left.

Photo: Carlos Carasas.

Hose Panning

This sign is done for the nozzlemen's view. The leader of the group advances to the front to signal them.

Photo: Carlos Carasas.

Change of Maneuver

This sign indicates a warning that the maneuver should change and will affect the group altogether.

Photo: Carlos Carasas.

Proceed

This sign should be done with enough time for assistant nozzleman (D, E) and others to prepare to move the hoses.

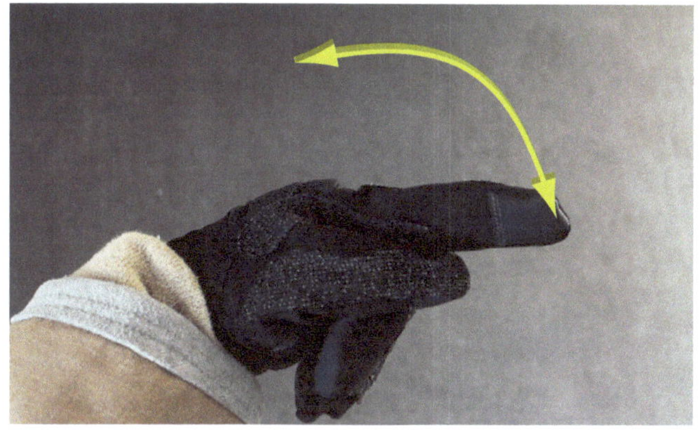

Photo: Carlos Carasas.

Proceed

This index finger is lowered to maintain the horizontal position, repetitive motion pointing towards the front meaning to proceed.

Photo: Carlos Carasas.

Something is Wrong

With this sign the leader is indicating to the group that something is not right or that the operation is no working. The next signal that could follow is an indicator for maneuver change.

Photo: Carlos Carasas.

Straight Stream

This sign indicates, nozzles should shoot a straight stream; the index finger must be drawn firm and high.

Photo: Carlos Carasas.

Narrow Fog

This sign indicates, nozzles should launch a narrow fog, also known as $15^0 - 45^0$.

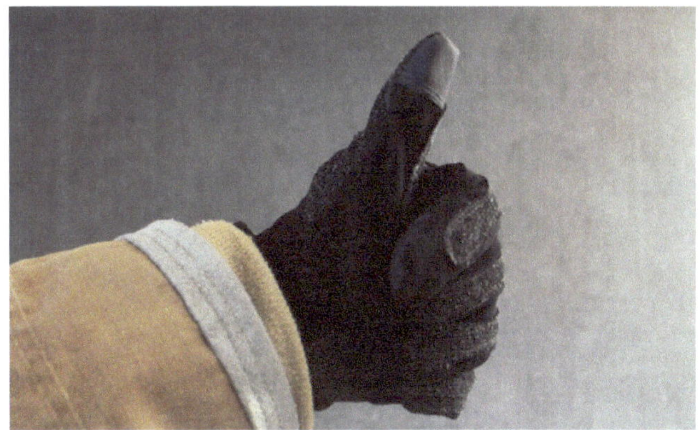

Photo: Carlos Carasas.

All Good

This sign indicates, the universal sign of success, it also means that the operation is being well executed and the extinction is occurring.

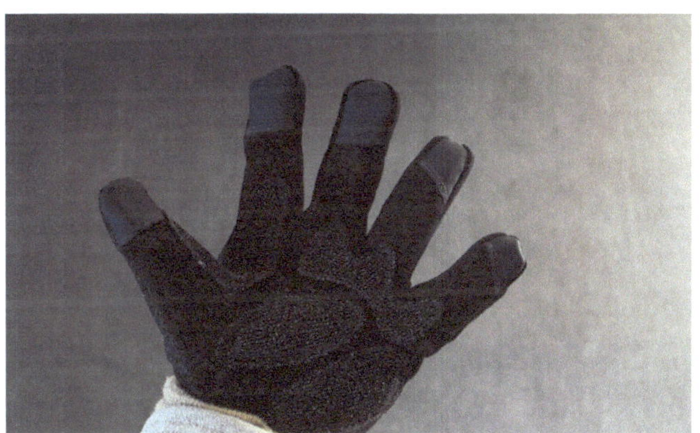

Photo: Carlos Carasas.

Wide fog

This sign indicates, nozzles should launch a wide-angle fog.

Photo: Carlos Carasas.

Stop or Pause
The stop signal is universal and should not be confused by indicating the need of the wide fog cloud which is characterized by having spread fingers.

The Author

José Musse is among the best sellers on fire protection literatures. With over 26 years of experience fighting fires; in 1996 he introduced the "Incident Command System" in Latin America. He also created a 3D virtual simulator for training firefighters, since 1998 Jose struggled to introduce to the United States, Australia and Latin America the concept of a giant waterbomber to deal with wildfires. He was Director of Operations, and Fire Chief of the consortium American, Canadian & Russian "Global Emergency Response", promoting the use of the Ilyushin—76 aircraft.

The United Nations Environment Program (UNEP) has cited his work on forest fires in Bolivia, as well as Syracuse University in New York. Recently a study by the University of Kent on emergency symbols has also mentioned his professional contribution in the field.

Currently he is Director of Fire Training Center of Peru. He lives between New York & Lima, Peru, active around the world lecturing and training firefighters as well as a consultant promoting new methods and systems, has written several technical manuals on fire tactics.

For 16 years Jose has been Director of **Desastres.org** online Magazine that aims to increase the operating efficiency of Hispanic firefighters as well as being an activist who has fought against corruption in the Emergency Services.